Nature's Wonders: Animals' Mass Migrations

Copyright Page

TITLE: Nature's Wonders: Animals' Mass Migrations

1ST Edition

Copyright @ 2023

Roberto M. Rodriguez. All rights reserved.

ISBN: 9798223144359

Table of Contents

Nature's Wonders: Animals' Mass Migrations

By Roberto Miguel Rodriguez

Chapter 1: Introduction to Animal Migrations

The Fascinating World of Animal Migrations

Welcome to the captivating realm of animal migrations, where creatures embark on incredible journeys across vast landscapes and unforgiving terrains. In this subchapter, we will delve into the wonders of nature's migrations, exploring the various habitats and species that participate in these awe-inspiring phenomena. Whether it's the vast oceans, arctic tundras, African savannahs, or dense rainforests, animals have adapted remarkable strategies to undertake their remarkable migrations.

Marine animal migrations take us into the depths of the world's oceans, where we will focus on the mass migrations of whales, dolphins, and other marine creatures. Discover how these magnificent creatures navigate vast distances, from their feeding grounds to their breeding grounds, and witness the challenges they face along the way.

Bird migrations, on the other hand, take us high up in the skies, where we will explore the incredible journeys of migratory birds across continents. From the Arctic tern's astonishing round-trip migration from the Arctic to the Antarctic, to the majestic flights of the bar-tailed godwits across the Pacific Ocean, we will uncover the secrets behind these avian adventurers.

Venturing into the insect world, we will delve into the fascinating world of insect migrations, such as the migration of Monarch butterflies. Witness the extraordinary transformation from caterpillar to butterfly, and follow their remarkable journey across North America, spanning several generations.

Moving to the Arctic, we will highlight the seasonal movements of Arctic animals like polar bears, walruses, and seals. Explore how these resilient creatures navigate the frozen landscapes in search of food and breeding opportunities, adapting to the harsh conditions of their environment.

Journeying to the African savannah, we will be captivated by the great migrations of wildebeests, zebras, and other animals across the vast grasslands. Witness the dramatic river crossings, the endless herds stretching across the horizon, and the intricate relationship between predator and prey.

Our exploration continues with river animal migrations, as we study the migration patterns of fish species, such as salmon, as they navigate rivers and waterways. Discover the remarkable instincts and adaptations that enable these fish to overcome countless obstacles and complete their life cycle.

In the arid deserts, we investigate how desert-dwelling animals like camels and gazelles adapt to their harsh environments during migration. Witness their incredible ability to endure extreme temperatures, scarce resources, and vast distances, as they traverse the unforgiving sands.

Moving to the grasslands, we explore the mass migrations of grazing animals like bison, antelope, and gazelles. Witness the awe-inspiring sight of thousands of animals on the move, following the lush pastures and avoiding predators in their quest for survival.

Diving into the dense rainforests, we discover the complex migrations of animals in these biodiverse habitats. From the tropical birds that traverse continents to find suitable breeding grounds, to the mammals that navigate through dense foliage, uncover the challenges and adaptations these creatures face.

Lastly, we examine high altitude animal migrations, where animals like yaks, ibex, and snow leopards adapt to their migrations in high altitude environments. Witness their incredible resilience to oxygen-deprived atmospheres and freezing temperatures as they traverse the rugged mountain landscapes.

In this subchapter, we will embark on a journey across the globe, unraveling the mysteries of animal migrations in various habitats. Prepare to be amazed by the resilience, instincts, and adaptations of these remarkable creatures as they undertake their incredible journeys across land, air, and sea.

Importance of Studying Animal Migrations

Understanding the importance of studying animal migrations is crucial for students interested in the fascinating world of wildlife. Animal migrations play a vital role in maintaining the delicate balance of ecosystems across various habitats around the world. By delving into the incredible journeys of different species, students can gain valuable insights into the interconnectedness of nature and the challenges these animals face during their migrations.

One of the significant benefits of studying animal migrations is the opportunity to learn about the adaptability and resilience of different species. For example, in the subchapter on marine animal migrations, students can explore the mass migrations of whales, dolphins, and other marine creatures. By understanding the patterns and drivers of these migrations, students can appreciate the incredible ability of these animals to navigate vast distances, find food sources, and reproduce in different parts of the ocean.

Similarly, studying bird migrations can provide students with a deeper understanding of the remarkable journeys undertaken by migratory birds across continents. From the Arctic to the African savannah, migratory birds face numerous obstacles, including weather conditions, predators, and habitat loss. By examining their migration routes and behaviors, students can develop an appreciation for the resilience and survival strategies of these avian species.

Insect migrations, such as the famous migration of Monarch butterflies, offer another intriguing area of study. By exploring the complexities of these migrations, students can learn about the delicate balance between these insects and their environment. They can also gain insights into the potential impacts of climate change and habitat destruction on these delicate creatures.

Studying animal migrations is not limited to specific habitats; it encompasses a wide range of ecosystems. From the Arctic to the rainforests, from deserts to grasslands, each habitat presents unique challenges and opportunities for migratory animals. By examining the migration patterns of different species in these diverse environments, students can develop a holistic understanding of how animals adapt to their surroundings and the importance of conserving these habitats for their survival.

Overall, studying animal migrations provides students with a window into the wonders of the natural world. It allows them to appreciate the intricate connections between species, habitats, and the environment. By understanding the challenges and adaptations of migratory animals, students can become advocates for conservation and contribute to the preservation of these incredible journeys for generations to come.

How Animal Migrations Are Studied

Understanding the intricacies of animal migrations is a fascinating field of study that requires a combination of scientific methods and technological advancements. Scientists and researchers employ several techniques to study the incredible journeys of wildlife, shedding light on the mysteries of these mass migrations.

One of the primary methods used to study animal migrations is satellite tracking. By attaching small GPS or radio transmitters to animals, researchers can monitor their movements in real-time. This technology has revolutionized our understanding of marine animal migrations, such as those of whales, dolphins, and other marine creatures. It has allowed scientists to track their migration routes, identify important feeding grounds, and monitor the impact of human activities on their populations.

For bird migrations, scientists use a combination of radar systems, bird banding, and geolocators. Radar systems help detect large flocks of birds in the sky, enabling researchers to estimate the number of birds and track their movement patterns. Bird banding involves placing small, numbered metal or plastic bands on birds' legs, allowing scientists to track individual birds and gather information about their migration routes and behavior. Geolocators are tiny devices that record light levels, helping scientists determine the locations birds visit during migration.

Insect migrations, such as the iconic migration of Monarch butterflies, are studied through a combination of citizen science efforts and tagging programs. Citizen scientists across North America help track the movements of Monarchs by reporting their sightings and tagging individual butterflies. These tags provide valuable information about the timing and routes of their migrations.

Arctic animal migrations, including those of polar bears, walruses, and seals, are studied using satellite imagery, aerial surveys, and on-the-ground observations. Scientists monitor the sea ice extent and its changes, which are critical for these animals' survival. They also use remote sensing techniques to track the movements of these animals across vast and often inaccessible regions.

Other techniques used to study animal migrations include acoustic monitoring, which involves recording the sounds animals make during migration, and DNA analysis, which helps identify the genetic connections between different populations.

By employing these diverse methods, scientists can unravel the mysteries behind the incredible journeys of animals across various ecosystems. The knowledge gained from these studies is essential for conservation efforts, as it helps identify critical habitats, migration corridors, and potential threats to these remarkable creatures. As students, understanding how animal migrations are studied opens up a world of possibilities for further research and exploration in the field of wildlife biology.

Chapter 2: Marine Animal Migrations

Introduction to Marine Animal Migrations

Welcome to the captivating world of marine animal migrations! In this subchapter, we will embark on a journey through the vast oceans to explore the incredible mass migrations of whales, dolphins, and other marine creatures. Prepare to be amazed by the remarkable feats of these majestic beings as they navigate the depths of the sea.

Marine animal migrations are a phenomenon that has fascinated scientists and nature enthusiasts for centuries. From the humpback whales' epic journey from the polar regions to the tropics, to the mesmerizing pods of dolphins leaping through the ocean waves, these migrations showcase the resilience and adaptability of marine life.

One of the most well-known marine migrations is the annual journey of the gray whales along the Pacific Coast. These magnificent creatures travel thousands of miles from their feeding grounds in the Arctic to their breeding grounds in warmer waters. Along the way, they face numerous challenges, including predators, changing ocean conditions, and human activities. Despite these obstacles, the gray whales persist in their journey, leaving us in awe of their determination.

But it's not just whales that undertake these incredible migrations. Dolphins, too, exhibit impressive movements across the oceans. From the charismatic bottlenose dolphins to the acrobatic spinner dolphins, these intelligent creatures travel in groups, or pods, searching for food, mating opportunities, and suitable habitats. Their synchronized swimming and playful antics make for a captivating sight, leaving us with a deeper appreciation for the wonders of marine life.

As we delve into the world of marine animal migrations, we will explore the factors that drive these extraordinary journeys. We will examine how the availability of food, temperature changes, breeding cycles, and instinctual behaviors shape the migration patterns of these creatures. Additionally, we will discuss the challenges they face, including climate change, pollution, and habitat destruction, and the conservation efforts being made to protect their migratory routes.

So, join us as we dive into the depths of the ocean and discover the awe-inspiring world of marine animal migrations. Through this subchapter, you will gain a newfound understanding of the intricate lives of these remarkable creatures and the importance of preserving their migratory routes for generations to come.

The Incredible Journeys of Whales

Whales, the gentle giants of the ocean, embark on some of the most extraordinary migrations on Earth. These magnificent marine creatures travel thousands of miles each year in search of food, mating grounds, and warmer waters. In this subchapter, we will dive into the fascinating world of whale migrations and explore the incredible journeys they undertake.

Whale migrations are a remarkable phenomenon that occurs across various species, including humpback whales, blue whales, and gray whales. These migrations can span vast distances, with some whales traveling from their feeding grounds in polar regions to their breeding grounds in tropical waters.

One of the most well-known whale migrations is that of the humpback whales. These majestic creatures undertake an annual journey from their feeding grounds in the cold waters of the Arctic or Antarctic to the warm breeding grounds in tropical regions. This migration can cover thousands of miles, and humpback whales are known to breach, sing, and engage in other spectacular behaviors along the way.

Another remarkable whale migration is that of the gray whales. These massive mammals undertake one of the longest migrations of any mammal, traveling from their summer feeding grounds in the Arctic to the warm lagoons of Baja California, Mexico, to give birth and mate. This journey spans around 12,000 miles round trip and is a true testament to the endurance and resilience of these incredible animals.

The reasons behind these epic migrations are multifaceted. Whales migrate to find abundant food sources, escape harsh weather conditions, and give birth in safer and warmer environments. The availability of food plays a crucial role in determining the routes and timing of these migrations.

Understanding whale migrations is essential for their conservation and protection. By studying their patterns and behaviors, scientists can gain insights into the health of their populations and the impacts of human activities on these magnificent creatures.

In conclusion, the incredible journeys of whales are a testament to the wonders of the natural world. Their migrations span thousands of miles and showcase the remarkable adaptability and resilience of these marine animals. By delving into the world of whale migrations, students can gain a deeper appreciation for the interconnectedness of ecosystems and the importance of preserving these incredible creatures for future generations.

Dolphins' Migratory Patterns

Dolphins are known for their intelligence, playful nature, and their impressive ability to navigate vast oceanic distances. In this subchapter, we will delve into the migratory patterns of dolphins and explore the incredible journeys they undertake.

Dolphins are highly social creatures that live in close-knit communities known as pods. These pods can consist of a few individuals or even hundreds of dolphins. One of the most fascinating aspects of dolphin migration is their ability to communicate and coordinate their movements with other members of their pod.

Many species of dolphins undertake seasonal migrations, traveling great distances in search of food, warmer waters, or suitable breeding grounds. These migrations can span hundreds or even thousands of miles, showcasing the remarkable endurance and adaptability of these marine mammals.

One well-known example of dolphin migration is the annual journey of the Pacific white-sided dolphins along the coast of North America. These dolphins travel from their summer feeding grounds in the colder waters of Alaska to the warmer waters of Southern California during the winter months. This migration allows them to follow their prey and avoid the harsh conditions of the colder regions.

Another remarkable migratory pattern is exhibited by the spinner dolphins in the Pacific Ocean. These dolphins travel from their daytime resting areas in the deep offshore waters to the shallower coastal areas at night to feed. This daily migration enables them to take advantage of the abundant food sources available near the coast.

Dolphin migrations are not only driven by the need for food but also by the desire to reproduce. Certain species, like the common dolphins, undertake long-distance migratory journeys to reach their preferred breeding grounds. These migrations involve large groups of dolphins swimming together in a synchronized fashion, creating an awe-inspiring spectacle.

Studying dolphin migratory patterns is crucial for understanding their behavior, population dynamics, and conservation needs. By tracking their movements and identifying their migration routes, scientists can gain valuable insights into the health of their habitats and the potential threats they face.

In conclusion, dolphins' migratory patterns are a testament to their adaptability and intelligence. Their ability to navigate vast distances and coordinate their movements with other members of their pod is truly remarkable. By exploring these migratory journeys, we can deepen our understanding of these incredible creatures and work towards their conservation and protection in the vast oceans they call home.

Other Remarkable Marine Creatures on the Move

While whales, dolphins, and other marine creatures are undoubtedly fascinating, there are many other remarkable species that undertake incredible journeys in the vast oceans. These migrations, although less known, are equally awe-inspiring and play a crucial role in maintaining the marine ecosystem's balance. In this chapter, we will dive into the captivating world of other marine creatures on the move.

One such creature is the leatherback sea turtle, renowned for its impressive migratory patterns. These ancient reptiles travel thousands of miles across oceans to reach their nesting grounds. Starting their journey from feeding areas in colder regions, they navigate through treacherous currents and navigate by using Earth's magnetic field. Witnessing the sight of a female leatherback laying her eggs on a remote beach is a true testament to the wonders of nature.

Another intriguing traveler is the remarkable Portuguese man o' war. Despite its appearance as a single organism, it is, in fact, a colony of specialized individuals working together. These floating colonies drift with the ocean currents, often covering vast distances in search of optimal feeding grounds. Their vibrant colors and venomous tentacles serve as a warning to potential predators, making them an intriguing subject of study.

The oceanic whitetip shark is yet another marine creature on the move. Known for their long-distance migrations, these apex predators traverse vast expanses of open ocean. They follow specific currents, often converging in areas rich in food sources, such as feeding frenzies around floating debris or aggregations of other marine animals. Their presence in these areas is essential for the health of the ecosystem, as they help regulate populations of other marine species.

Lastly, we cannot overlook the incredible journeys of the graceful manta rays. These gentle giants undertake epic migrations, often spanning thousands of miles, in search of plankton-rich waters. Their migration patterns are influenced by seasonal changes, food availability, and the need to mate and give birth. Observing a school of mantas gliding effortlessly through the ocean is a sight that leaves a lasting impression on any marine enthusiast.

In conclusion, while the mass migrations of whales and dolphins may steal the spotlight, there are numerous other marine creatures embarking on extraordinary journeys. The leatherback sea turtle, Portuguese man o' war, oceanic whitetip shark, and manta rays are just a few examples of the diverse and captivating migrations taking place beneath the ocean's surface. By studying and understanding these remarkable creatures, we can gain a deeper appreciation for the wonders of the marine world and the intricate connections that sustain life in our oceans.

Chapter 3: Bird Migrations

The Wonders of Bird Migrations

Introduction:

Bird migrations are one of the most awe-inspiring phenomena in the natural world. Every year, millions of birds embark on incredible journeys across continents, defying all odds and showcasing the marvels of nature. In this subchapter, we will explore the fascinating world of bird migrations, understanding the reasons behind their journeys, the challenges they face, and the extraordinary adaptations that enable them to complete these incredible feats.

The Beauty of Bird Migrations:

Bird migrations are a testament to the indomitable spirit of nature. From the elegant Arctic Tern, which travels an astounding 44,000 miles each year, to the majestic Ruby-throated Hummingbird, which crosses the Gulf of Mexico non-stop, birds traverse vast distances, crossing mountains, oceans, and deserts. Witnessing the synchronized flights of thousands of birds is a sight that leaves both scientists and nature enthusiasts in awe.

Reasons for Migration:

Birds undertake long and perilous journeys for several reasons. The most common motivation is the search for better food and breeding grounds. As seasons change, birds follow the availability of food, moving from colder regions to warmer ones where insects and fruits are abundant. Additionally, the need for suitable nesting sites and favorable climates drives birds to migrate.

Navigation and Orientation:

The ability of birds to navigate accurately over thousands of miles is nothing short of astonishing. They use a variety of cues, including the position of the sun, stars, magnetic fields, and even landmarks, to guide their journeys. Some birds possess an innate sense of direction, while others rely on learning from older individuals or using celestial cues.

Challenges and Adaptations:

Bird migrations present numerous challenges that birds must overcome. They face extreme weather conditions, exhaustion, predation, and the risk of colliding with buildings or communication towers. To tackle these obstacles, birds have evolved remarkable adaptations. They store energy in the form of fat, grow thicker feathers for insulation, and possess enlarged hearts and lungs to sustain their endurance flights.

Conservation and Importance:

Understanding bird migrations is crucial for their conservation. Habitat loss, climate change, and human activities pose significant threats to the survival of migratory birds. By studying their migration patterns, we can identify key stopover sites, breeding grounds, and wintering areas, allowing us to implement effective conservation measures and protect these incredible journeys for future generations.

Conclusion:

Bird migrations are a testament to the resilience and adaptability of nature. The journeys undertaken by migratory birds are nothing short of marvelous, captivating the imagination of students and nature enthusiasts alike. By delving into the wonders of bird migrations, we gain a deeper appreciation for the intricate connections between species and ecosystems, highlighting the need for their conservation and the preservation of the natural world.

Long-Distance Migratory Birds

In the vast realm of animal migrations, few journeys are as awe-inspiring and captivating as those undertaken by long-distance migratory birds. These incredible avian travelers embark on epic journeys across continents, traversing vast distances in search of favorable breeding grounds, abundant food sources, and hospitable climates. In this subchapter, we will delve into the fascinating world of long-distance migratory birds and explore the remarkable adaptations that allow them to accomplish these extraordinary feats.

One of the most remarkable aspects of long-distance migratory birds is their ability to navigate over thousands of kilometers with astonishing precision. Through an intricate combination of celestial cues, landmarks, and environmental signals, these birds are able to chart their course and reach their intended destinations with remarkable accuracy. Students will discover how factors such as the position of the sun, magnetic fields, and even the stars play a crucial role in guiding these avian travelers.

Additionally, students will learn about the incredible physical and physiological adaptations that enable long-distance migratory birds to undertake these grueling journeys. From their lightweight bodies and streamlined shapes that facilitate efficient flight to their enhanced endurance and navigational abilities, these birds have evolved a remarkable set of traits that allow them to conquer vast distances.

Furthermore, this subchapter will highlight some of the most iconic long-distance migratory birds, such as the Arctic tern, the bar-tailed godwit, and the ruby-throated hummingbird. Students will discover the fascinating details of their migratory patterns, including their breeding grounds, wintering locations, and the challenges they face along the way.

Lastly, this subchapter will explore the importance of conservation efforts in protecting the habitats and ecosystems that are crucial for the survival of long-distance migratory birds. By understanding the threats these birds face, such as habitat loss, climate change, and human interference, students will gain a deeper appreciation for the need to preserve these incredible animals and their remarkable journeys.

In conclusion, the world of long-distance migratory birds is a testament to the wonders of nature and the resilience of these remarkable creatures. By studying their journeys, adaptations, and the challenges they face, students will develop a greater understanding and appreciation for the extraordinary journeys undertaken by long-distance migratory birds.

Songbirds' Seasonal Journeys

As the seasons change and the weather shifts, a remarkable phenomenon takes place in the world of songbirds. These small, colorful creatures embark on incredible journeys across continents, navigating vast distances to find the perfect breeding grounds and feeding areas. In this subchapter, we will delve into the fascinating world of songbird migrations and discover the secrets behind their seasonal journeys.

Songbirds, also known as passerines, are a diverse group of birds that belong to the order Passeriformes. They come in all shapes, sizes, and colors, from the vibrant orioles to the melodious warblers. What sets songbirds apart is their ability to produce complex songs, which they use for communication and courtship.

One of the most remarkable aspects of songbirds is their incredible navigation skills. Using a combination of celestial cues, landmarks, and magnetic fields, they are able to navigate across vast distances with astonishing accuracy. It is believed that they can even detect subtle changes in the Earth's magnetic field, allowing them to make adjustments to their flight paths as needed.

The timing of songbird migrations is closely linked to the availability of resources. As the days become shorter and temperatures drop, songbirds start their long journeys to warmer climates where food is abundant.

They rely on a variety of food sources, including insects, fruits, and nectar, which they find along their migration routes.

During their migrations, songbirds face numerous challenges and obstacles. They must navigate unfamiliar landscapes, avoid predators, and overcome fatigue and hunger. Many songbirds undertake non-stop flights lasting several days, covering thousands of miles without any rest. It is truly a testament to their endurance and determination.

In addition to the physical challenges, songbirds also face the threat of habitat loss and climate change. As human activities continue to impact the environment, the habitats that songbirds rely on for breeding and feeding are being destroyed or altered. This puts their populations at risk and underscores the importance of conservation efforts to protect these remarkable creatures.

In conclusion, the seasonal journeys of songbirds are a testament to their remarkable abilities and the wonders of the natural world. By understanding and appreciating these incredible migrations, we can gain a deeper appreciation for the interconnectedness of ecosystems and the importance of preserving the habitats that these beautiful birds rely on.

Birds of Prey and Their Migrations

Birds of prey, also known as raptors, are a majestic group of birds known for their incredible hunting skills and aerial prowess. But did you know that many of these birds also undertake remarkable migrations across continents? In this subchapter, we will delve into the world of bird migrations and explore the incredible journeys of these magnificent creatures.

Bird migrations are truly a marvel of nature. Every year, millions of birds take to the skies and embark on long and arduous journeys in search of food, breeding grounds, and favorable climates. These migrations can span thousands of miles and involve crossing vast oceans, mountains, and deserts.

One of the most well-known migratory birds of prey is the osprey. These magnificent birds travel from their breeding grounds in North America to wintering grounds as far south as South America. Along their journey, ospreys face numerous challenges, including navigating treacherous weather conditions and finding enough food to sustain themselves.

Another fascinating migratory raptor is the peregrine falcon. Known for its incredible speed and agility, the peregrine falcon undertakes one of the longest migrations of any bird. These birds breed in the Arctic tundra and then travel south to wintering grounds in South America. Along the way, they navigate through various habitats, including grasslands, deserts, and rainforests.

Birds of prey rely on a variety of navigational cues to guide them during their migrations. They use landmarks, such as coastlines and mountain ranges, as well as celestial cues, like the position of the sun and stars, to orient themselves. Some raptors also possess a remarkable ability to sense the Earth's magnetic field, which helps them navigate accurately over long distances.

Understanding the patterns and challenges of bird migrations is crucial for their conservation and management. Climate change, habitat loss, and human activities pose significant threats to these magnificent creatures. By studying their migrations, scientists can gain insights into their population dynamics, habitat requirements, and potential conservation strategies.

In the following chapters, we will explore other fascinating animal migrations, such as those of marine creatures, insects, Arctic animals, African savannah wildlife, river fish species, desert-dwelling animals, grassland grazers, rainforest inhabitants, and high altitude species. Each migration is unique and offers valuable insights into the adaptability and resilience of these remarkable animals.

So, join us on this journey through the world of animal migrations. From the skies to the oceans, from the deserts to the rainforests, we will uncover the wonders of nature's incredible journeys.

Chapter 4: Insect Migrations

Understanding Insect Migrations

In the vast world of animal migrations, one of the most intriguing and often overlooked journeys is that of insects. From the delicate flutter of a butterfly's wings to the synchronized movements of swarming locusts, the migration of insects is a fascinating phenomenon that deserves our attention. In this subchapter, we will delve into the incredible world of insect migrations, with a special focus on the migration of Monarch butterflies.

Insects, despite their small size, are capable of embarking on long and perilous journeys, sometimes spanning thousands of miles. They migrate for various reasons, including escaping unfavorable conditions, finding better resources, or even reproducing. These migrations often involve millions, if not billions, of individuals, creating awe-inspiring spectacles in the natural world.

One of the most famous insect migrations is that of the Monarch butterflies. Every year, millions of Monarchs embark on an epic journey from their summer breeding grounds in North America to their wintering grounds in Mexico. This journey spans an astonishing 3,000 miles and takes several generations of butterflies to complete. The Monarchs navigate using a combination of innate instincts and environmental cues, such as the position of the sun and the Earth's magnetic field.

But what drives these tiny creatures to undertake such a challenging journey? The answer lies in the availability of resources. Monarchs rely heavily on milkweed plants for their survival, as their caterpillars feed exclusively on these plants. However, as the seasons change, the availability of milkweed diminishes in North America, forcing the butterflies to migrate south in search of better food sources and warmer climates.

Understanding the intricacies of insect migrations is not only fascinating but also crucial for our understanding of the natural world. By studying these migrations, scientists can gain valuable insights into the ecological processes that shape our planet. For example, the migration of certain insects can have profound impacts on ecosystems, such as pollinating plants or serving as a vital food source for other animals.

In conclusion, the world of insect migrations is a captivating one, filled with incredible journeys and extraordinary adaptations. By exploring the migration of Monarch butterflies and other insect species, we can gain a deeper appreciation for the complexity and interconnectedness of the natural world. So, let us embark on this enthralling journey into the fascinating world of insect migrations and discover the wonders that await us.

The Monarch Butterfly's Epic Journey

Deep within the realm of insect migrations lies a truly remarkable journey - the epic migration of the Monarch butterfly. This delicate creature embarks on an incredible odyssey, spanning thousands of miles, capturing the imagination of scientists and nature enthusiasts alike.

Every year, millions of Monarch butterflies take flight from their breeding grounds in North America to their overwintering sites in Mexico. This awe-inspiring journey is a testament to the resilience and adaptability of these tiny insects.

The journey begins in the late summer and early autumn, as the Monarchs sense the changing seasons and prepare for their long flight. Fueling themselves on nectar from flowering plants, they build up the necessary energy reserves to sustain them on their arduous journey. As temperatures drop and days grow shorter, the Monarchs take to the skies, forming vast swarms that darken the horizon.

Navigating through a complex network of air currents and wind patterns, the Monarchs embark on a migration that spans several generations. The first generation of butterflies, born during the migration, will continue the journey, guided by an innate sense of direction and the position of the sun.

Though the Monarchs cover great distances, their path remains remarkably consistent. They follow a well-worn route, known as the Monarch Highway, which takes them across the United States and into Mexico. Along the way, they encounter numerous obstacles, including unpredictable weather conditions, predators, and habitat loss. However, their determination and adaptability enable them to overcome these challenges.

Once they reach their overwintering sites in the mountain forests of Mexico, the Monarchs cluster together in vast colonies, creating a spectacle that is truly awe-inspiring. Here, they hibernate, conserving their energy until the arrival of spring signals the time to begin their return journey.

The Monarchs' migration holds great significance not only for the species itself but also for the ecosystems they inhabit. As they travel, they play a vital role as pollinators, aiding in the reproduction of countless plant species. Their presence also serves as an indicator of the health of their habitats, as their declining numbers serve as a warning of environmental degradation.

The Monarch butterfly's epic journey is a testament to the astonishing capabilities of nature's wonders. By studying and understanding these migrations, we gain insight into the delicate balance and interconnectedness of the natural world. As students of nature's wonders, we have the privilege of witnessing these incredible journeys and preserving them for generations to come.

Other Intriguing Insect Migrations

In addition to the captivating migration journeys of whales, birds, and other animals, the world of insects is also brimming with fascinating migrations. These tiny creatures embark on extraordinary journeys, often spanning great distances and defying the odds. One such remarkable migration is that of the Monarch butterflies.

Every year, millions of Monarch butterflies undertake an incredible migration from Canada and the United States to their wintering grounds in Mexico. This journey covers a staggering distance of over 3,000 miles. What makes this migration even more intriguing is that it spans multiple generations of butterflies. The journey begins with the monarchs that hatch in the northern regions, and as they mate and reproduce along the way, their offspring continue the migration, eventually reaching the final destination in Mexico.

But Monarch butterflies are not the only insects that embark on extraordinary migrations. Another notable example is the dragonfly. These agile creatures can cover hundreds or even thousands of miles during their annual migrations. Some dragonfly species undertake these impressive journeys in search of better breeding grounds or to escape unfavorable weather conditions. They navigate using the sun, landmarks, and even the Earth's magnetic field.

Insect migrations are not limited to just butterflies and dragonflies. Many other insects, such as beetles, moths, and even certain species of ants, undertake seasonal migrations. These journeys are driven by various factors, including the search for food, favorable breeding conditions, or the need to escape harsh weather.

Understanding insect migrations is crucial for scientists and ecologists as it provides valuable insights into ecosystem dynamics and the interconnectedness of different habitats. It also helps in understanding the impact of climate change on these delicate creatures and the ecosystems they inhabit.

For students interested in the wonders of nature, exploring the world of insect migrations offers a unique perspective. It sheds light on the incredible resilience and adaptability of these tiny creatures and teaches us valuable lessons about the interconnectedness of all living beings.

So, dive into the fascinating world of insect migrations and discover the untold stories of these incredible journeys. From the delicate beauty of Monarch butterflies to the tenacity of dragonflies, there is much to learn and marvel at. By delving into the intricacies of insect migrations, students can gain a deeper appreciation for the wonders of the natural world and the diverse strategies employed by different species to survive and thrive.

Chapter 5: Arctic Animal Migrations

Seasonal Movements of Arctic Animals

The Arctic is home to some of the most incredible and resilient creatures on the planet. From polar bears to walruses and seals, these animals have adapted to survive in one of the harshest environments on Earth. One of the most fascinating aspects of Arctic wildlife is their seasonal movements, which are driven by the changing seasons and the availability of food.

During the summer months, the Arctic is a bustling ecosystem. The sea ice melts, creating open water and allowing marine animals to thrive. This is the time when many Arctic species, such as seals and walruses, gather in large groups to breed and give birth. The newborns are carefully nurtured by their mothers until they are strong enough to venture out on their own.

As fall approaches and the temperatures drop, the sea ice begins to form again. This signals the start of the Arctic winter and triggers the migration of many animals. One of the most iconic Arctic migrations is that of the polar bear. These magnificent creatures rely on the sea ice to hunt for seals, their primary source of food. As the ice expands, polar bears embark on a journey in search of suitable hunting grounds. They can travel hundreds of miles, navigating treacherous ice floes and enduring harsh weather conditions.

Similarly, Arctic birds, such as the Arctic tern, undertake incredible migrations. These small birds fly from the Arctic to the Antarctic and back, covering a distance of over 40,000 miles each year. They follow the summer from pole to pole, taking advantage of the abundance of food in both hemispheres.

The seasonal movements of Arctic animals are not only driven by the search for food but also by the need to find suitable habitats for breeding and raising their young. The Arctic may be a harsh environment, but it is also a place of incredible beauty and diversity. By understanding and appreciating the seasonal movements of Arctic animals, we can gain a deeper appreciation for their remarkable adaptations and the delicate balance of nature in this unique ecosystem.

So, join us as we delve into the world of Arctic animal migrations and explore the awe-inspiring journeys of polar bears, walruses, and seals. Discover how these incredible creatures navigate the changing seasons and overcome the challenges of their icy home. Get ready to be amazed by the resilience and strength of Arctic wildlife and gain a new appreciation for the wonders of nature.

Polar Bears: The Masters of Arctic Migration

The Arctic region is home to some of the most extreme conditions on Earth, and yet it is also a place of great beauty and diversity. One of the most iconic species that call the Arctic home is the polar bear. These magnificent creatures have adapted to thrive in the harsh Arctic environment, and their annual migration is a testament to their resilience and survival instincts.

Polar bears are known for their ability to traverse vast distances in search of food and suitable breeding grounds. Their migration patterns are closely tied to the movement of sea ice, which serves as their hunting platform and a means of travel. As the seasons change and the ice begins to melt, polar bears must move to new areas to find seals, their primary source of food.

The migration of polar bears is a complex and fascinating process. It begins in the early spring when the bears emerge from their winter dens and start their journey across the Arctic. They rely on their incredible sense of smell to detect seals hidden beneath the ice, and they can travel up to hundreds of miles in search of a suitable hunting ground.

During their migration, polar bears face numerous challenges, including the threat of starvation and encounters with other predators. They must navigate treacherous ice floes, swim long distances in freezing waters, and endure extreme weather conditions. Yet, they have evolved to adapt to these challenges, with a thick layer of blubber to keep them warm and powerful limbs to help them swim.

The migration of polar bears is not just a physical journey but also a vital part of their reproductive cycle. Female polar bears will often travel to specific areas to give birth and raise their cubs. These areas, known as maternity dens, provide a safe haven for the vulnerable cubs during their first months of life.

As students of nature's wonders, it is important to understand and appreciate the incredible journeys of animals like polar bears. Their migration is a testament to their adaptability and resilience in the face of a rapidly changing environment. By studying and conserving their habitats, we can ensure the survival of these magnificent creatures and the delicate balance of the Arctic ecosystem.

In the next chapter, we will delve into the fascinating world of African savannah migrations, where wildebeests, zebras, and other animals embark on epic journeys across the vast grasslands of Africa. Stay tuned!

Walruses and Seals on the Move

In the icy waters of the Arctic, a remarkable journey begins every year as walruses and seals embark on their epic migrations. These marine creatures navigate vast distances, braving treacherous conditions to find food, mates, and suitable birthing grounds. Join us as we delve into the fascinating world of Arctic animal migrations and explore the incredible journeys of these magnificent creatures.

Walruses, known for their distinctive tusks and large bodies, undertake seasonal migrations in search of food. During the summer months, as the ice melts and their preferred prey, such as clams and mollusks, becomes scarce, walruses travel to new feeding grounds. These migrations can take them hundreds of miles, as they follow the receding ice and navigate through chilly waters. Witnessing these massive animals on the move is a sight to behold, as they form sprawling herds and travel together, their calls echoing through the Arctic air.

Seals, on the other hand, undertake migrations for different reasons. Some species, like the ringed seals, journey to find suitable areas for giving birth and raising their pups. These seals rely on stable ice floes, where they excavate snow caves to protect their young from predators and the harsh Arctic weather. It is an incredible feat of adaptation, as these seals must time their migrations perfectly to ensure the availability of suitable ice conditions.

The migrations of walruses and seals highlight the delicate balance of life in the Arctic ecosystem. As these animals travel, they play a crucial role in dispersing nutrients and shaping the distribution of marine life. They also face numerous challenges, including shrinking sea ice, changing ocean currents, and increased human activity. Understanding their movements and the factors that influence them is vital for their conservation and the preservation of the Arctic environment.

In this chapter, we will explore the intricacies of walrus and seal migrations, their adaptations to life in the Arctic, and the importance of these journeys for the overall health of the ecosystem. Through captivating stories, stunning photographs, and engaging activities, we will learn about the challenges these animals face and the conservation efforts being made to protect their habitats. Get ready to dive into the icy depths of the Arctic and witness the awe-inspiring migrations of walruses and seals.

Chapter 6: African Savannah Migrations

The Great Migrations of the African Savannah

Welcome to the fascinating world of animal migrations! In this subchapter, we will delve into the extraordinary journeys of wildlife in the African savannah. The African savannah is home to some of the most iconic and awe-inspiring mass migrations on the planet.

One of the most renowned migrations takes place in the Serengeti, where millions of wildebeests, zebras, and other animals embark on an epic journey in search of food and water. This incredible spectacle is known as the Great Migration. Every year, these animals follow the rains, traveling thousands of kilometers across the vast savannah.

The migration begins in the southern Serengeti, where the wildebeests give birth to their young. As the dry season approaches, the animals instinctively start moving northward, following the fresh grasses and water sources. Along the way, they face numerous challenges, including treacherous river crossings and encounters with predators like lions and crocodiles.

The migration not only impacts the animals themselves but also has a profound effect on the entire ecosystem. It helps to distribute nutrients across the landscape, fertilizing the soil and supporting the growth of diverse plant species. Additionally, the wildebeests and zebras create pathways through the grasslands, which other animals can then utilize.

The Great Migration is a prime example of how animals have adapted to their environment and developed intricate strategies for survival. By moving in massive herds, they increase their chances of evading predators and finding food. They also rely on their collective intelligence to navigate the ever-changing landscape.

For students interested in the African savannah and its incredible wildlife, studying these migrations provides a window into the complex web of life in this unique ecosystem. It allows us to appreciate the interconnectedness of species and the delicate balance required for their survival.

In the following chapters, we will explore other remarkable mass migrations, including those of marine creatures, birds, insects, Arctic animals, river species, desert dwellers, grassland grazers, rainforest inhabitants, and high-altitude explorers. Each migration presents its own set of challenges and adaptations, showcasing the incredible diversity and resilience of nature's wonders.

So, join us on this journey as we unravel the mysteries of animal migrations and discover the extraordinary feats of strength, endurance, and instinct that allow these creatures to traverse vast distances in search of a better life.

Wildebeests: The Iconic Migrants

In the vast African savannah, one phenomenon captures the attention of wildlife enthusiasts and scientists alike - the great migrations of wildebeests. Every year, millions of these iconic animals embark on an incredible journey, traversing vast distances in search of greener pastures. This subchapter explores the captivating world of wildebeest migrations, shedding light on their behaviors, challenges, and the significance of these mass movements.

The wildebeest migration is a sight to behold, as thousands upon thousands of these creatures move together in synchronized harmony. Their journey spans across the Serengeti in Tanzania and the Maasai Mara in Kenya, covering over 1,800 miles. This epic migration is driven by the pursuit of fresh grazing lands and the need to survive the dry seasons.

The migration begins in the Serengeti, where wildebeests give birth to their young. The abundance of food and water attracts predators such as lions, hyenas, and crocodiles, making it a perilous journey right from the start. Yet, the wildebeests must press on, and as the dry season approaches, they start their northward migration towards the Maasai Mara.

Crossing treacherous rivers, battling strong currents and lurking predators, the wildebeests face countless challenges during their migration. The most perilous river crossing occurs at the Mara River, where crocodiles lie in wait for their next meal. Only the strongest and luckiest survive this dangerous crossing, while others fall victim to the jaws of death.

The significance of wildebeest migrations extends beyond the survival of these remarkable creatures. Their movements shape entire ecosystems, as they serve as a vital food source for predators and contribute to the dispersal of nutrients across vast landscapes. Moreover, the wildebeest migration attracts tourists from around the world, providing economic benefits to local communities and supporting conservation efforts.

Studying the wildebeest migration allows us to gain insights into the complex dynamics of ecosystems and the delicate balance of nature. By understanding the challenges faced by these animals, we can develop strategies to protect and conserve their habitats. Furthermore, the wildebeest migration serves as a reminder of the interconnectedness of all living beings and the importance of preserving the natural world.

In conclusion, the wildebeest migration is a testament to the wonders of nature and the incredible journeys undertaken by animals. This subchapter provides a glimpse into the world of wildebeest migrations, allowing students to appreciate the beauty, challenges, and significance of these iconic migrations. Through studying the wildebeests' incredible journey, we can deepen our understanding of the natural world and work towards its preservation for generations to come.

Zebras and Other Animals on the Move

In the vast African savannah, an awe-inspiring spectacle takes place every year – the great migrations of wildebeests, zebras, and other animals. These incredible journeys are an integral part of the natural world, showcasing the resilience and adaptability of wildlife.

In this subchapter, we will delve into the African savannah migrations, focusing on the iconic zebras and their fellow travelers. We will explore the reasons behind these mass movements, the challenges faced by these animals, and the remarkable strategies they employ to survive.

The African savannah is a dynamic ecosystem that experiences dramatic changes in food availability and water sources throughout the year. Zebras and other grazers rely on the seasonal rains to replenish the vegetation, which is essential for their survival. As the dry season approaches, these animals must venture in search of greener pastures and water, often traveling vast distances.

One of the most renowned migrations is the Serengeti wildebeest migration, where millions of wildebeests, zebras, and gazelles make their way across the plains. These animals form massive herds, creating a mesmerizing sight as they move in unison. The zebras play a vital role in the migration, as their keen sense of smell helps guide the herd to water sources.

During these migrations, animals face numerous challenges, including predators, river crossings, and limited resources. We will explore how zebras and other species navigate these obstacles, employing strategies such as safety in numbers, communication through vocalizations and body language, and an innate sense of direction.

Additionally, we will examine the impact of these migrations on the African savannah ecosystem. The movement of large herds helps rejuvenate the land, spreading seeds and fertilizing the soil. We will discuss the symbiotic relationships between zebras, wildebeests, and other animals, as well as their interactions with predators like lions and hyenas.

By studying the migrations of zebras and other animals in the African savannah, we gain a deeper understanding of the intricate web of life and the remarkable adaptations that enable these creatures to thrive. We will also draw comparisons to other mass migrations, such as those in the marine, bird, insect, arctic, river, desert, grassland, rainforest, and high altitude environments.

Join us on this journey as we explore the wonders of nature's mass migrations, uncovering the secrets of these incredible journeys and gaining a newfound appreciation for the resilience and beauty of the animal kingdom.

Chapter 7: River Animal Migrations

Navigating the Rivers: Fish Migration Patterns

In the vast world of animal migrations, one of the most fascinating journeys takes place beneath the surface of our rivers and waterways. Fish, such as salmon, embark on incredible migrations, navigating through treacherous currents and obstacles to reach their ultimate destinations. In this chapter, we will dive into the world of river animal migrations and explore the patterns and challenges these fish face along the way.

Fish migrations in rivers are crucial for their lifecycle and survival. They undertake these journeys to find suitable spawning grounds, feed, or seek refuge from changing environmental conditions. One of the most well-known river migrations is that of salmon. These remarkable fish are born in freshwater rivers, then migrate to the ocean where they spend a significant portion of their lives before returning to their natal rivers to reproduce.

The journey of a salmon is nothing short of miraculous. They can navigate hundreds or even thousands of miles, overcoming countless obstacles like waterfalls, dams, and predators. The ability of these fish to remember the scent of their home river is a remarkable adaptation that helps guide them back to the exact spot where they were born.

Salmon are not the only fish species that undertake such incredible migrations. Many other freshwater fish, such as eels and sturgeons, also navigate rivers and waterways to complete their lifecycles. Each species faces unique challenges along their migratory routes, and scientists are constantly studying their patterns and behaviors to better understand and protect these valuable ecosystems.

In this chapter, we will explore the different types of fish migrations, the challenges they face, and the importance of maintaining healthy river ecosystems for their survival. We will also delve into the role of humans in the conservation and management of these migratory fish, including the construction of fish ladders and the removal of barriers to facilitate their journeys.

By understanding the intricate world of fish migrations in rivers, we can gain a deeper appreciation for the delicate balance of nature and the interconnectedness of ecosystems. Join us on this journey as we dive into the fascinating world of river animal migrations and discover the wonders that lie beneath the surface.

The Extraordinary Journey of Salmon

In the vast world of animal migrations, few journeys are as remarkable as that of the salmon. These incredible fish undertake an extraordinary journey, navigating rivers and defying the odds to complete their life cycle. Join us as we dive into the captivating world of salmon migration.

Salmon are born in freshwater rivers, but their destiny lies in the vastness of the ocean. As young fry, they make their way downstream, facing numerous obstacles along the way. From treacherous rapids to hungry predators, their survival is constantly at stake. Only the strongest and luckiest individuals manage to make it to the open sea.

Once in the ocean, the salmon embark on a remarkable feeding frenzy. They grow rapidly, feasting on abundant marine resources. As they mature, they undergo incredible physiological changes, preparing themselves for the next stage of their journey.

After spending several years in the ocean, the salmon's instinct compels them to return to the exact river where they were born. This homing ability is one of nature's most astounding phenomena. Guided by their amazing sense of smell, the salmon navigate hundreds, and sometimes thousands, of miles back to their birthplace.

The return journey is no less arduous than their initial migration. Upstream, against strong currents, and leaping over waterfalls, the salmon face countless obstacles. Many don't survive the grueling journey, but those that do arrive at their spawning grounds, driven by an innate desire to reproduce.

In a breathtaking display of determination, the female salmon dig nests in the riverbed, known as redds, where they lay their eggs. The males fertilize the eggs, and the cycle begins anew. Exhausted and depleted, the salmon ultimately succumb to their efforts, completing the circle of life.

The journey of the salmon is not only awe-inspiring, but it also plays a crucial role in maintaining the balance of ecosystems. As they return to the rivers, they bring vital nutrients from the ocean, enriching the surrounding habitats and supporting a diverse array of life.

In conclusion, the extraordinary journey of salmon is a testament to the resilience and adaptability of nature's wonders. From their humble beginnings in freshwater rivers to their epic odyssey in the ocean and back, salmon captivate our imagination and remind us of the incredible journeys undertaken by wildlife.

Other Fascinating River-Dwelling Species on the Move

While we often associate mass migrations with land and marine animals, it is important not to overlook the incredible journeys undertaken by river-dwelling species. These animals face unique challenges as they navigate rivers and waterways, adapting to changing water levels and currents. In this subchapter, we will explore the fascinating world of river animal migrations and the species that call these water bodies their home.

One of the most well-known river migrations is that of the salmon. Every year, millions of salmon make the arduous journey from the ocean back to their freshwater spawning grounds. This incredible feat involves swimming upstream against powerful currents, leaping over waterfalls, and overcoming numerous obstacles along the way. Not only is this migration essential for the survival of the salmon population, but it also has far-reaching effects on the ecosystem, as the nutrients brought by the returning salmon contribute to the health of the surrounding habitat.

But salmon are not the only river-dwelling species on the move. Many other fish species undertake remarkable migrations, such as eels and sturgeon. Eels, for example, migrate from freshwater rivers to the ocean to spawn, while sturgeon undertake long journeys to reach their spawning grounds. These migrations are crucial for the reproduction and survival of these species, as they ensure the mixing of genetic diversity and the continuation of their populations.

In addition to fish, there are other fascinating river-dwelling species that migrate as well. River otters, for instance, are known to travel long distances along river systems in search of food and suitable habitats. These playful and agile creatures rely on healthy river ecosystems to support their migrations and sustain their populations.

Furthermore, certain bird species rely on rivers as important stopover sites during their migrations. For example, the American Dipper, a small aquatic songbird, is known for its unique ability to dive and walk underwater in search of food. These birds migrate along river corridors, utilizing the abundant resources found in these habitats.

In conclusion, river animal migrations are a vital component of our natural world. From the impressive journeys of salmon to the agile movements of river otters, these migrations play a crucial role in maintaining healthy ecosystems and sustaining biodiversity. By studying and understanding these fascinating river-dwelling species, we can gain a deeper appreciation for the wonders of nature and the interconnectedness of all living creatures.

Chapter 8: Desert Animal Migrations

Surviving the Desert: Animal Migrations in Harsh Environments

Deserts are known for their extreme temperatures, scorching heat, and lack of water. Yet, even in these harsh environments, remarkable animal migrations take place. In this subchapter, we will delve into the fascinating world of desert animal migrations and explore how desert-dwelling creatures such as camels and gazelles adapt to their challenging surroundings during these incredible journeys.

The desert poses numerous challenges for animals during migration. The scarcity of water and food presents a constant struggle for survival. However, through thousands of years of adaptation, desert animals have developed unique strategies to overcome these obstacles.

One of the most iconic desert migrants is the camel. Equipped with specialized physiological features, camels can travel for long distances in search of water and vegetation. Their humps store fat, not water as commonly believed, which they can utilize when food and water sources are scarce. Additionally, their long legs help them navigate through the desert sands, and their broad feet provide stability and prevent them from sinking into the hot sand.

Gazelles, on the other hand, have evolved to survive in the desert by being incredibly efficient with water consumption. They can obtain the moisture they need from the vegetation they eat and can go for long periods without drinking water. Their light bodies and long limbs enable them to move quickly across the desert, avoiding predators and seeking out better feeding grounds.

During migration, desert animals must also contend with extreme temperatures. To cope with the scorching heat, many desert creatures are nocturnal, venturing out to feed and travel during the cooler nighttime hours. This strategy helps them conserve energy and avoid the risk of dehydration.

In this subchapter, we will explore the migratory patterns of various desert-dwelling animals, such as the Arabian oryx, the addax, and the sand gazelle. We will analyze their routes, timing, and adaptations that allow them to survive in the desert. We will also examine the role of human intervention in protecting these fragile ecosystems and the challenges faced by desert migrants due to habitat loss and climate change.

Join us as we uncover the secrets of desert animal migrations, and gain a deeper understanding of how these incredible creatures navigate and thrive in one of the world's harshest environments.

Camels: The Desert Navigators

The desert is a harsh and unforgiving environment, with scorching heat, shifting sands, and little to no water. Yet, amidst this seemingly inhospitable landscape, one animal stands tall and thrives: the camel. Known as the "ship of the desert," camels are exceptional navigators that have adapted to survive and thrive in these extreme conditions.

Camels are well-suited for desert life due to their unique physiological features. Their humps, for example, are not filled with water as commonly believed, but with fat reserves that provide a crucial energy source during long journeys without food. This adaptation allows camels to go without drinking water for extended periods, making them ideal for life in arid regions.

Another remarkable feature of camels is their ability to withstand extreme temperatures. Their long legs help them navigate through hot sand, while their broad, tough feet protect them from the scorching ground. Additionally, their nostrils can be closed to keep out sand and dust during sandstorms, ensuring their survival even in the harshest of conditions.

Camels are also excellent navigators, possessing an innate sense of direction that allows them to travel vast distances across the desert. They have been used as transportation and guides by desert-dwelling communities for centuries. Their keen sense of smell helps them locate water sources, even from miles away, ensuring their survival during long migrations.

During the migration, camels form social groups called caravans, which consist of several individuals traveling together. This not only provides protection against predators but also allows for cooperation in finding food and water. The leader of the caravan, usually an experienced adult camel, guides the group using their exceptional navigational skills.

Camel migrations are essential for the survival of these magnificent creatures. As they move across the desert, they help disperse seeds, ensuring the growth and survival of plants in these barren landscapes. In turn, these plants provide food and shelter for other desert animals, creating a delicate balance within the ecosystem.

The migration of camels is a testament to their remarkable adaptability and resilience. As students of nature, we can learn valuable lessons from these desert navigators, such as the importance of adaptation and cooperation in overcoming challenges. Camels remind us that even in the harshest of environments, life finds a way to thrive, and nature's wonders never cease to amaze us.

Gazelles and Other Desert-Dwelling Migrants

In the vast and unforgiving deserts of the world, a remarkable phenomenon takes place - the migration of desert-dwelling animals. Among these incredible migrants are the graceful gazelles, known for their speed and agility in the harsh desert environment. This subchapter will delve into the fascinating world of gazelles and other desert-dwelling migrants, exploring how they adapt to their harsh surroundings and undertake epic journeys in search of food, water, and safety.

The story of gazelle migration is a testament to the resilience and adaptability of these remarkable creatures. In the scorching heat of the desert, where resources are scarce and temperatures can reach extreme highs, gazelles have evolved unique strategies to survive. They are well adapted to conserve water, with efficient kidneys and the ability to obtain moisture from the plants they eat. Their long legs and light bodies enable them to traverse the sandy terrain with ease, while their keen senses help them detect predators from a distance.

During the dry season, when food and water become scarce, gazelles embark on long and arduous migrations in search of greener pastures. These migrations can cover vast distances, with gazelles traveling hundreds of miles in search of more favorable conditions. Along the way, they must navigate treacherous terrain, including towering sand dunes and rocky outcrops. Despite the challenges they face, gazelles are able to maintain their strength and stamina, thanks to their efficient metabolism and ability to adapt to limited resources.

But gazelles are not the only desert-dwelling migrants. Other animals, such as camels and desert foxes, also undertake remarkable journeys in search of food and water. Camels, known as the "ships of the desert," are well adapted to the harsh conditions, with the ability to store water in their humps and withstand extreme temperatures. Desert foxes, on the other hand, have developed the ability to go for long periods without water, relying on their adaptations to the arid environment.

Understanding the migrations of desert-dwelling animals provides valuable insights into the delicate balance of life in these harsh environments. It highlights the interconnectedness of different species and the importance of adaptation in the face of adversity. By studying these incredible journeys, students can gain a deeper appreciation for the wonders of nature and the resilience of the animal kingdom.

So, join us on a journey to the desert, where gazelles and other desert-dwelling migrants navigate the unforgiving sands in search of sustenance and survival. Discover the secrets of their adaptations, the challenges they face, and the marvels of nature that unfold along the way.

Chapter 9: Grassland Animal Migrations

The Mass Migrations of Grazing Animals

In the vast grasslands of the world, a breathtaking phenomenon occurs each year - the mass migrations of grazing animals. These incredible journeys are a testament to the remarkable adaptability and resilience of these animals. In this subchapter, we will delve into the fascinating world of grassland animal migrations, exploring the reasons behind their movements and the challenges they face along the way.

Grassland ecosystems, with their abundance of nutrient-rich grasses, provide the perfect habitat for grazing animals such as bison, antelope, and gazelles. However, the availability of food is not constant, and as seasons change, so do the grazing opportunities. In response to this, these animals have evolved to undertake long and arduous migrations in search of better grazing grounds.

One of the most famous grassland migrations takes place in the Serengeti, where millions of wildebeests and zebras embark on an epic journey across the African savannah. This incredible spectacle, known as the Great Migration, is driven by the need for fresh pastures and water sources. Students will be captivated by the sheer numbers and the collaborative nature of this migration, as different species work together to navigate the challenges of the journey.

But it's not just in Africa where these mass migrations occur. In North America, herds of bison undertake impressive journeys across the Great Plains, following the seasonal growth of grasses. These massive animals, once numbering in the millions, now face the challenges of habitat loss and fragmentation, making their migrations all the more important for their survival.

Throughout this subchapter, students will learn about the various adaptations that enable these grazing animals to undertake their migrations. From their powerful legs and endurance to their keen senses and innate navigational abilities, these animals have honed their skills over generations.

Additionally, students will gain insight into the ecological significance of these migrations. These journeys not only ensure the survival of the animals themselves but also have far-reaching impacts on the grassland ecosystems they traverse. By spreading seeds, fertilizing the soil, and controlling vegetation, these animals play a vital role in maintaining the health and diversity of their habitats.

The mass migrations of grazing animals are truly awe-inspiring, showcasing the wonders of the natural world. Through this subchapter, students will gain a deeper appreciation for the challenges these animals face and the remarkable adaptations that enable them to undertake these incredible journeys. As they dive into the world of grassland animal migrations, students will develop a greater understanding of the delicate balance between animals and their environments, inspiring them to become advocates for the conservation of these magnificent creatures.

Bison: The Great Wanderers

In the vast grasslands of North America, a magnificent creature roams freely, embodying the spirit of true wanderers. The bison, scientifically known as Bison bison, is a majestic symbol of strength, resilience, and the vastness of the natural world. In this subchapter, we will delve into the incredible mass migrations of bison, exploring their fascinating journeys across the grassland ecosystems.

The grasslands, also known as prairies, serve as the primary habitat for bison. These expansive landscapes, stretching as far as the eye can see, provide an abundance of grasses and other vegetation that sustain the bison's massive herds. However, the availability of resources fluctuates with the seasons, prompting the bison to embark on epic migrations in search of greener pastures.

During the spring and summer months, bison herds graze on the rich grasses of the northern prairies, feasting upon the nutritious vegetation that fuels their growth and reproduction. As the seasons change and the temperatures drop, the bison sense the approaching winter and begin their southward migration, following ancient paths ingrained in their collective memory.

The southern grasslands offer the bison a refuge from the harsh winter conditions, as the region experiences milder temperatures and less snowfall. Here, the bison find sheltered valleys and open plains where they can continue to graze on the sparse grasses that manage to survive the colder months. The bison's adaptation to these extreme environments is truly remarkable, as they endure freezing temperatures and use their thick fur and layers of fat to stay warm.

As the arrival of spring signals the onset of new growth, the bison herds reverse their migration, heading back north to the fertile grazing grounds of the northern prairies. This cyclic pattern of movement ensures the survival of the species, as it allows the bison to exploit the seasonal abundance of resources across their vast range.

The mass migrations of bison not only showcase their ability to adapt to changing environments but also contribute to the overall health and balance of the grassland ecosystems. Their grazing behavior helps maintain the diversity of plant species and stimulates new growth, benefiting countless other organisms that depend on the prairies for their survival.

In conclusion, the bison's mass migrations are a testament to the incredible journeys undertaken by wildlife across the globe. Their story of resilience, adaptation, and the pursuit of greener pastures serves as a captivating example of nature's wonders. By understanding and appreciating these migrations, we can gain a deeper appreciation for the interconnectedness of all living beings and the importance of preserving their habitats for generations to come.

Antelope and Gazelles: Fleet-Footed Migrants

In the vast grasslands of the world, a remarkable phenomenon takes place - the mass migrations of antelope and gazelles. These fleet-footed creatures embark on incredible journeys, covering vast distances in search of food, water, and suitable breeding grounds. In this subchapter, we will delve into the world of antelope and gazelles, exploring their adaptations, migration patterns, and the challenges they face along the way.

Antelope and gazelles are well-known for their speed and agility. With their slender bodies and powerful legs, they are built for swift movements across the open plains. During migration, these animals can reach speeds of up to 60 miles per hour, allowing them to cover great distances in a short amount of time. This speed is crucial for their survival, as it helps them evade predators and find food and water sources.

The migration patterns of antelope and gazelles are often influenced by seasonal changes in their environment. As the dry season sets in and water sources become scarce, these animals gather in large herds and embark on long journeys in search of greener pastures and permanent water sources. This journey can span hundreds of miles, requiring tremendous strength and endurance.

One of the most famous migrations of antelope is the Serengeti migration in East Africa. Here, millions of wildebeests, zebras, and gazelles travel in a circular route, following the rains and fresh grasses.

This migration not only provides them with food but also plays a vital role in the ecosystem, as their movements help to fertilize the soil and promote the growth of new vegetation.

During their migration, antelope and gazelles face numerous challenges. They must navigate treacherous terrains, cross rivers infested with crocodiles, and evade predators such as lions, cheetahs, and hyenas. The journey is not without its dangers, and many individuals do not survive. However, those that do reach their destination are rewarded with abundant resources and the opportunity to reproduce, ensuring the survival of their species.

In conclusion, the mass migrations of antelope and gazelles are truly awe-inspiring. These fleet-footed migrants showcase the incredible adaptability and resilience of animals in the face of changing environments. By studying their migration patterns and understanding their unique adaptations, we can gain a deeper appreciation for the wonders of the natural world and the interconnectedness of all living beings.

Chapter 10: Rainforest Animal Migrations

The Complex Migrations of Rainforest Animals

Rainforests are some of the most diverse and complex ecosystems on Earth, and the migrations of animals within these habitats are equally fascinating. In this chapter, we will delve into the intricate world of rainforest animal migrations, exploring the incredible journeys undertaken by tropical birds and mammals.

One of the most iconic rainforest migrations is that of the tropical birds. These colorful creatures undertake long-distance flights across vast distances, following specific routes known as flyways. These flyways are determined by factors such as food availability, climate, and breeding grounds. For example, the Scarlet Macaws of Central and South America travel hundreds of kilometers in search of fruiting trees, while the Arctic Tern migrates from the Arctic to the Antarctic and back, covering a staggering 70,900 kilometers each year.

Mammals in rainforest environments also engage in unique migratory behaviors. One example is the movement of large herbivores, such as tapirs and peccaries, in search of food and water. They traverse the dense vegetation, facing numerous obstacles and predators along the way. Additionally, certain rainforest mammals, such as bats and primates, undertake seasonal migrations to find suitable habitats for breeding and foraging.

The complex migrations of rainforest animals are not only driven by the need for resources but also play a crucial role in maintaining the health and balance of these ecosystems. Through their movements, animals disperse seeds, pollinate plants, and provide a source of food for predators. Without these migrations, the rainforest ecosystem would suffer greatly.

However, rainforest animal migrations face numerous challenges and threats. Deforestation, habitat fragmentation, and climate change all pose significant risks to the survival of these migratory species. As students, it is important for us to understand the impacts of human activities on rainforest ecosystems and work towards their conservation and protection.

In the following chapters, we will explore other remarkable animal migrations in different ecosystems, ranging from the marine world to the African savannah. By studying these incredible journeys, we can gain a deeper appreciation for the wonders of nature and the interconnectedness of all living beings.

Tropical Birds: Migratory Wonders

In the vast world of animal migrations, few migrations are as captivating and awe-inspiring as those of tropical birds. These beautiful creatures embark on incredible journeys across continents, defying boundaries and pushing the limits of their endurance. In this subchapter, we will delve into the extraordinary lives of tropical birds and unravel the mysteries of their migratory wonders.

Tropical birds, with their vibrant plumage and melodious songs, captivate our hearts and stir our imagination. But what truly sets them apart is their ability to traverse vast distances in search of favorable breeding grounds and abundant food sources. From the colorful Scarlet Macaw to the majestic Swallow-tailed Kite, these migratory wonders undertake journeys that span thousands of miles, showcasing their remarkable navigational skills and adaptability.

One of the most remarkable examples of tropical bird migration is the journey of the Arctic Tern. This small bird, weighing only a few ounces, embarks on an astonishing round-trip migration from the Arctic to the Antarctic, covering a mind-boggling distance of over 40,000 miles. Imagine traveling from the North Pole to the South Pole and back, all in one year! This incredible feat is made possible by the bird's ability to harness wind patterns and navigate with pinpoint accuracy using celestial cues.

But the Arctic Tern is not alone in its migratory prowess. Many other tropical birds, such as the Ruby-throated Hummingbird and the Blackpoll Warbler, undertake epic journeys that span continents. These birds brave treacherous weather conditions, cross vast oceans, and navigate through unfamiliar landscapes to reach their destination. Their migrations are a testament to their survival instincts and their remarkable ability to adapt to a changing environment.

The reasons behind tropical bird migrations are multifaceted. Some birds migrate to find suitable breeding grounds, while others follow the seasonal availability of food resources. The diverse habitats that these birds inhabit, ranging from rainforests to savannahs and from mangroves to coastal areas, provide them with ample opportunities for survival and reproduction.

In conclusion, tropical birds are truly migratory wonders. Their ability to navigate across vast distances, their adaptability to different environments, and their innate survival instincts make them a fascinating subject of study. By understanding and appreciating the journeys of these incredible creatures, we can gain a deeper appreciation for the wonders of nature and the intricate web of life that connects us all. So, let us embark on a journey of discovery and explore the extraordinary world of tropical bird migrations.

Mammals on the Move in Rainforest Habitats

In the dense and vibrant world of rainforest habitats, a remarkable phenomenon takes place - the migration of mammals. Rainforests are home to some of the most diverse and unique species on the planet, and their movements within these lush ecosystems are nothing short of extraordinary. In this subchapter, we will explore the captivating journeys of mammals in rainforest habitats, shedding light on their behaviors, adaptations, and the challenges they face.

One of the most iconic rainforest migrants is the orangutan. These majestic creatures traverse the treetops of Southeast Asia in search of food and suitable nesting sites. With their long arms and impressive agility, orangutans swing effortlessly from branch to branch, covering vast distances as they forage for fruits, leaves, and insects. Their migrations play a crucial role in dispersing seeds, contributing to the regeneration of the rainforest.

Another fascinating rainforest migrant is the jaguar. These powerful and elusive predators roam the dense undergrowth of Central and South American rainforests. Jaguars are known to travel long distances in search of prey, navigating through the intricate network of trees, rivers, and other obstacles. Their migrations help maintain a balanced ecosystem by controlling the populations of herbivores and preventing overgrazing.

The rainforest is also a haven for primates, including monkeys and apes. Among them, spider monkeys stand out for their incredible agility and ability to cover large distances. These acrobatic creatures swing from tree to tree using their prehensile tails, allowing them to move quickly and efficiently within their forest home. Their migrations are influenced by the availability of fruits and the need to find suitable mates.

Additionally, some smaller rainforest mammals, such as bats and rodents, undertake seasonal migrations to escape the harsh dry seasons or find better food sources. These creatures may travel long distances, often unnoticed, as they seek out more favorable conditions in different parts of the rainforest.

The complex web of mammal migrations in rainforest habitats is a testament to the adaptability and resilience of these remarkable creatures. However, they face numerous challenges, including habitat loss, deforestation, and fragmentation, which threaten their survival and disrupt the delicate balance of the rainforest ecosystem.

By studying and understanding the movements of mammals in rainforest habitats, we can gain valuable insights into the importance of preserving these environments and ensuring the survival of the species that call them home. Let us marvel at the wonders of rainforest animal migrations and work towards their conservation, so that future generations can continue to witness these incredible journeys of wildlife.

Chapter 11: High Altitude Animal Migrations

Thriving in High Altitude Environments: Animal Adaptations

In the awe-inspiring world of animal migrations, few journeys are as challenging and remarkable as those undertaken in high altitude environments. From the towering peaks of the Himalayas to the rugged landscapes of the Andes, animals like yaks, ibex, and snow leopards have evolved unique adaptations to navigate and thrive in these extreme conditions.

One of the most crucial adaptations for high altitude animals is their ability to cope with low oxygen levels. At such elevations, the air becomes thin and oxygen becomes scarce. To combat this, animals have developed larger lungs and more efficient respiratory systems, enabling them to extract as much oxygen as possible from each breath. In addition, their blood contains a higher concentration of red blood cells, which helps to transport oxygen more effectively throughout their bodies.

Another key adaptation is their ability to regulate body temperature in freezing conditions. Many high altitude animals have developed thick fur or feathers to provide insulation against the cold. This not only keeps them warm but also helps to reduce heat loss. Some animals, like the snow leopard, even have a dense layer of fat beneath their fur to provide extra warmth during the harsh winter months.

These animals have also adapted their behavior and feeding habits to the unique challenges of high altitude environments. For example, yaks have evolved to graze on tough, fibrous grasses that grow at high altitudes. Their strong jaws and specialized teeth allow them to break down these coarse plants and extract nutrients efficiently. Similarly, ibex have adapted to climb steep, rocky slopes in search of food, using their strong hooves and agile bodies to navigate treacherous terrain.

In addition to these physical adaptations, high altitude animals have also developed specialized senses to survive in their harsh environments. For example, the snow leopard has incredible eyesight, allowing it to spot prey from great distances across the vast, snowy landscapes. The ibex, on the other hand, has excellent balance and coordination, enabling it to navigate narrow ledges and cliff faces with ease.

Thriving in high altitude environments is no easy feat, but these incredible animals have mastered the art of adaptation. Through their unique physiological, behavioral, and sensory adaptations, they are able to not only survive but thrive in some of the most inhospitable places on Earth. Studying their journeys and adaptations is not only fascinating but also provides invaluable insights into the resilience and adaptability of the natural world.

Yaks: High Altitude Nomads

In the vast expanse of high altitude environments, where freezing temperatures and thin air make survival a constant challenge, there exists a remarkable creature known as the yak. These sturdy and resilient animals have adapted to life in some of the harshest conditions on Earth and play a vital role in the ecosystems of the Himalayan region.

Yaks are native to the mountainous regions of Central Asia, including Tibet, Nepal, and Mongolia. They are well-suited to high altitude living, with a thick coat of long, shaggy hair that provides insulation against the cold. Their large lungs and strong hearts allow them to efficiently extract oxygen from the thin mountain air, enabling them to navigate steep slopes and climb to altitudes of over 5,000 meters.

One of the most fascinating aspects of yaks is their annual migration patterns. As the seasons change and the weather becomes increasingly challenging, yaks embark on journeys in search of food and better grazing grounds. During the summer months, when the high alpine meadows are lush and abundant, yaks ascend to the highest peaks, taking advantage of the abundant vegetation.

However, as winter approaches and the temperatures drop, the yaks must descend to lower altitudes in search of more favorable conditions. They move in herds, often led by a dominant female, known as the "nak" or matriarch. These migrations can cover vast distances, with yaks traveling for days or even weeks to reach their winter grazing grounds.

The adaptability of yaks is not limited to their physical characteristics and migration patterns. They have also played an important role in the lives of local communities for centuries. Yaks are highly valued for their milk, meat, and wool, which provide essential resources for the people living in these remote regions. They are also used as pack animals, carrying heavy loads across treacherous mountain trails.

In conclusion, yaks are truly remarkable creatures that have successfully carved out a niche in some of the most extreme environments on the planet. Their ability to migrate between high and low altitudes in search of food and better conditions is a testament to their resilience and adaptability. By studying the migration patterns of yaks, we can gain a deeper understanding of how animals thrive in high altitude environments and the delicate balance of ecosystems in these regions.

Ibex and Snow Leopards: Mastering High Altitude Migrations

At the highest peaks of the world, where the air is thin and the temperatures are freezing, a unique group of animals have mastered the art of high altitude migrations. Among them are the Ibex and the Snow Leopards, two remarkable species that navigate the treacherous terrain of the mountains with skill and precision.

The Ibex, a species of wild goat, is found in the mountainous regions of Europe, Asia, and Africa. These agile creatures are built for climbing, with sturdy hooves and a muscular body. During the summer months, when the snow begins to melt and the grasses grow, the Ibex leaves its winter habitat and embarks on a journey to higher altitudes in search of fresh food.

The Ibex's migration is not without its challenges. As they ascend to higher elevations, they must navigate steep slopes and rocky cliffs, using their powerful legs to leap from ledge to ledge. They rely on their keen senses of sight and hearing to detect any potential predators lurking nearby, such as the elusive Snow Leopard.

The Snow Leopard is a true master of high altitude survival. This elusive and majestic creature is perfectly adapted to its harsh environment. With its thick fur and large paws, it can withstand the freezing temperatures and move effortlessly through the snow-covered peaks. The Snow Leopard's migration is driven by the availability of prey, such as the Ibex and other mountain-dwelling animals.

During the winter months, when food becomes scarce in the higher elevations, the Snow Leopard descends to lower altitudes in search of easier hunting grounds. This descent can be perilous, as the steep slopes and icy conditions make it difficult to maintain balance. However, the Snow Leopard's powerful muscles and keen sense of balance allow it to navigate these treacherous slopes with ease.

Both the Ibex and the Snow Leopard have not only adapted to the extreme conditions of high altitude environments but have also mastered the art of migration. Their incredible journeys are a testament to the resilience and adaptability of nature's wonders.

In conclusion, the high altitude migrations of the Ibex and Snow Leopards are a fascinating example of how animals have adapted to the challenges of their environment. As students of nature's wonders, we can learn valuable lessons from these remarkable species about the importance of resilience, adaptation, and the delicate balance of ecosystems in extreme habitats.

Chapter 12: Conclusion and Reflection

Recap of the Incredible Journeys Explored

In the captivating world of wildlife, there are countless stories of incredible journeys taken by various animals across the globe. Throughout this book, "Nature's Wonders: Animals' Mass Migrations: A Student's Guide to the Incredible Journeys of Wildlife," we have explored the remarkable migrations of different species, each adapted to their unique environments and challenges. Let's recap some of the most awe-inspiring journeys we have encountered.

We began by diving into the depths of the ocean, where we discovered the mass migrations of marine creatures, particularly whales, dolphins, and other majestic beings. These marine animals undertake long-distance travels in search of food, breeding grounds, or better climate conditions. Witnessing their immense power and grace as they traverse vast expanses of water was truly humbling.

Moving to the skies, we marveled at the migratory patterns of birds that span continents. From the Arctic Tern's incredible journey from the Arctic to the Antarctic and back, to the iconic spectacle of the great wildebeest migration across the African savannah, these avian travelers demonstrated unparalleled determination and resilience.

Venturing into the small but mighty world of insects, we were fascinated by the migration of Monarch butterflies, which journey thousands of miles to reach their overwintering sites in Mexico. These delicate creatures showcased the remarkable abilities of even the tiniest of beings to navigate vast distances.

Heading to the Arctic, we explored the seasonal movements of Arctic animals such as polar bears, walruses, and seals. These remarkable creatures have adapted to survive in one of the harshest environments on Earth, undertaking perilous journeys in search of food and breeding opportunities.

Our exploration of river animal migrations introduced us to the incredible odysseys of fish species, most notably salmon, as they navigate rivers and waterways. Witnessing their determination to overcome treacherous obstacles and complete their life cycles was nothing short of extraordinary.

In the arid landscapes of deserts, we investigated how animals like camels and gazelles adapt to their harsh environments during migration. Their ability to endure extreme conditions and find sustenance in barren lands was a testament to their remarkable resilience.

Moving to grasslands, we marveled at the mass migrations of grazing animals like bison, antelope, and gazelles. These herds traverse vast expanses of grasslands, following seasonal changes and seeking fresh pastures. Their synchronized movements were a sight to behold.

In the lush rainforests, we dived into the complex migrations of animals such as tropical birds and mammals. The intricate web of life in these biodiverse habitats revealed the interconnectedness of species and the delicate balance essential for their survival.

Lastly, we ventured to high altitudes, where animals like yaks, ibex, and snow leopards adapt to their migrations in extreme environments. These resilient creatures have evolved unique physiological and behavioral traits to thrive in the thin air and freezing temperatures of mountainous regions.

Throughout this journey, we have witnessed the extraordinary lengths to which animals go to ensure their survival and perpetuation of their species. From the depths of the ocean to the highest peaks, nature never ceases to amaze. So, let us continue our exploration of the incredible journeys of wildlife and discover the wonders that await us in the natural world.

The Importance of Protecting and Preserving Animal Migrations

Introduction:

Welcome to the subchapter on the importance of protecting and preserving animal migrations. In this section, we will delve into the significance of these incredible journeys and why it is crucial for us as students and nature enthusiasts to understand and protect them.

Why are Animal Migrations Important?

Animal migrations are awe-inspiring natural phenomena that have been occurring for centuries. They play a vital role in maintaining the balance and health of various ecosystems around the world. Let's explore some key reasons why animal migrations are important.

1. Biodiversity and Ecosystem Health:

Animal migrations contribute to the biodiversity of different ecosystems. These movements allow for the transfer of nutrients, seeds, and genetic diversity, ensuring the survival and adaptation of various species. By protecting animal migrations, we are safeguarding the overall health and resilience of ecosystems.

2. Pollination and Seed Dispersal:

Migratory animals, such as birds and insects, play a critical role in pollination and seed dispersal. They transfer pollen between plants, enabling the reproduction of numerous plant species. Additionally, animals aid in seed dispersal, helping plants colonize new areas and maintain healthy populations.

3. Nutrient Cycling:

Migratory animals also contribute to nutrient cycling. For example, the annual migration of salmon brings essential nutrients from the ocean to freshwater ecosystems. The carcasses of salmon provide vital nutrients to other animals and plants, fueling the productivity of these ecosystems.

4. Economic Importance:

Animal migrations hold significant economic value. Many wildlife-based industries, such as ecotourism and wildlife photography, rely on the presence of migratory animals. By protecting these migrations, we support local economies and promote sustainable development.

5. Cultural and Educational Value:

Migratory animals have cultural importance for many communities around the world. They feature prominently in folklore, art, and traditional practices. Furthermore, studying animal migrations provides invaluable educational opportunities for students like us, fostering a deeper understanding and appreciation for the natural world.

Conclusion:

Understanding and protecting animal migrations is vital for the well-being of ecosystems, biodiversity, and human societies. By recognizing the importance of these incredible journeys, we can advocate for their preservation and contribute to the conservation efforts that ensure the survival of migratory species. Let us embrace the wonders of nature's migrations and work towards a future where these incredible journeys continue to inspire and benefit us all.

Inspiring Future Research and Conservation Efforts

As students, we have embarked on an incredible journey to explore the mass migrations of various animal species across the globe. Throughout our exploration, we have witnessed the awe-inspiring journeys of wildlife, from the depths of the ocean to the highest peaks of the mountains. These extraordinary migrations have not only captivated our minds but have also ignited a passion within us to contribute to the research and conservation efforts of these amazing creatures.

The study of nature's wonders, such as animals' mass migrations, opens up a vast realm of possibilities for future research. By delving deeper into the mysteries behind these migrations, we can uncover invaluable insights about the ecological significance and adaptations of these species. For instance, in our exploration of marine animal migrations, we have learned about the incredible journeys of whales, dolphins, and other marine creatures. This knowledge can be utilized to develop effective conservation strategies and marine protected areas to safeguard their habitats.

Moreover, our understanding of bird migrations across continents can help us recognize the importance of preserving stopover sites and flyways, which are crucial for their survival. By studying the intricate navigation systems employed by migratory birds, researchers can uncover the secrets of their incredible journey and potentially apply this knowledge to enhance navigation technologies for human use.

Insect migrations, such as the migration of Monarch butterflies, offer another fascinating area of research. By studying the behavior, physiology, and genetics of these insects, we can gain insight into their remarkable ability to navigate across vast distances. This knowledge can contribute to the conservation of these delicate creatures and their habitats.

The migratory patterns of animals in the Arctic, African savannah, rivers, deserts, grasslands, rainforests, and high altitude environments present unique research opportunities. By studying the challenges faced by animals during migration in these diverse ecosystems, we can gain valuable knowledge about their adaptations and the impacts of climate change on their survival.

Our exploration of nature's wonders has left us with a sense of responsibility to protect and conserve these incredible migrations. By inspiring future research and conservation efforts, we can ensure the preservation of these animals' habitats and their ability to undertake these awe-inspiring journeys for generations to come. As students, we have the power to make a difference by advocating for the protection of these species and their habitats, supporting organizations and initiatives dedicated to their conservation, and pursuing careers in research and conservation biology. Together, we can contribute to a future where these incredible migrations continue to inspire and amaze.